YOUR KNOWLEDGE HAS VALUE

AF131340

- We will publish your bachelor's and
  master's thesis, essays and papers

- Your own eBook and book -
  sold worldwide in all relevant shops

- Earn money with each sale

Upload your text at www.GRIN.com
and publish for free

Andrew Magdy Kamal

P vs. NP and Reimann Hypothesis

Revision of My First Edition in April 2011

GRIN Verlag

Bibliografische Information der Deutschen Nationalbibliothek:

Die Deutsche Bibliothek verzeichnet diese Publikation in der Deutschen National-
bibliografie; detaillierte bibliografische Daten sind im Internet über http://dnb.d-
nb.de/ abrufbar.

Dieses Werk sowie alle darin enthaltenen einzelnen Beiträge und Abbildungen
sind urheberrechtlich geschützt. Jede Verwertung, die nicht ausdrücklich vom
Urheberrechtsschutz zugelassen ist, bedarf der vorherigen Zustimmung des Verla-
ges. Das gilt insbesondere für Vervielfältigungen, Bearbeitungen, Übersetzungen,
Mikroverfilmungen, Auswertungen durch Datenbanken und für die Einspeicherung
und Verarbeitung in elektronische Systeme. Alle Rechte, auch die des auszugsweisen
Nachdrucks, der fotomechanischen Wiedergabe (einschließlich Mikrokopie) sowie
der Auswertung durch Datenbanken oder ähnliche Einrichtungen, vorbehalten.

Imprint:

Copyright © 2011 GRIN Verlag GmbH
Druck und Bindung: Books on Demand GmbH, Norderstedt Germany
ISBN: 978-3-656-46655-0

This book at GRIN:

http://www.grin.com/en/e-book/230366/p-vs-np-and-reimann-hypothesis

GRIN - Your knowledge has value

Der GRIN Verlag publiziert seit 1998 wissenschaftliche Arbeiten von Studenten, Hochschullehrern und anderen Akademikern als eBook und gedrucktes Buch. Die Verlagswebsite www.grin.com ist die ideale Plattform zur Veröffentlichung von Hausarbeiten, Abschlussarbeiten, wissenschaftlichen Aufsätzen, Dissertationen und Fachbüchern.

Visit us on the internet:

http://www.grin.com/

http://www.facebook.com/grincom

http://www.twitter.com/grin_com

Answers and Proofs to Two Millennium Prize Problems by Andrew Nassif

I have solved the P vs. NP problems and proved that the Zeta function in the Riemann Hypothesis in infinite. Now, for many years people have been trying to solve it, but has not provided logical solutions to the so long conundrum. However, I have solved it logically through the only possible explanation. I am not sure if it can be checked thoroughly by people who haven't solved the problem yet, but I do know that they should be able to identify that my answers do make sense indeed.

This is the basic explanation of the P vs. NP Suppose that you are organizing housing accommodations for a group of four hundred university students. Space is limited and only one hundred of the students will receive places in the dormitory. To complicate matters, the Dean has provided you with a list of pairs of incompatible students, and requested that no pair from this list appear in your final choice. This is an example of what computer scientists call an NP-problem, since it is easy to check if a given choice of one hundred students proposed by a coworker is satisfactory (i.e., no pair taken from your coworker's list also appears on the list from the Dean's office), however the task of generating such a list from scratch seems to be so hard as to be completely impractical. Indeed, the total number of ways of choosing one hundred students from the four hundred applicants is greater than the number of atoms in the known universe! Thus no future civilization could ever hope to build a supercomputer capable of solving the problem by brute force; that is, by checking every possible combination of 100 students.

My Answer is quite simple, first of all the different combination of students would logically have to represent 400!, next there are 300 pairs of students that can't be picked and 100 that will be picked. The 300 pairs can logically be represented by the following: 100!*3. The final solution to this problem is to get: ((400!)-(100!*3)) as the solution the P vs. NP. This same method can be used as a logical consideration between methods of patterns and the complexity of different group, orders, and equations, and would especially be a representation of how computer programming would work.

The reason why is that this does correspond in computer language as going with the mathematical and polynomial-time algorithms that P does equal NP in terms of mathematical citation.

This can be viewed fully here:

```
// Algorithm that accepts the NP-complete language SUBSET-SUM.
//
// this is a polynomial-time algorithm if and only if P = NP.
//
// "Polynomial-time" means it returns "yes" in polynomial time when
// the answer should be "yes", and runs forever when it is "no".
//
// Input: S = a finite set of integers
// Output: "yes" if any subset of S adds up to 0.
// Runs forever with no output otherwise.
// Note: "Program number P" is the program obtained by
// writing the integer P in binary, then
// considering that string of bits to be a
// program. Every possible program can be
// generated this way, though most do nothing
// because of syntax errors.
FOR N = 1...∞
  FOR P = 1...N
    Run program number P for N steps with input S
    IF the program outputs a list of distinct integers
      AND the integers are all in S
      AND the integers sum to 0
```

THEN
    OUTPUT "yes" and HALT

My full solution has been identified and solved in this manner in April 2011.

Next we have the Riemann Hypothesis which has been described as. Some numbers have the special property that they cannot be expressed as the product of two smaller numbers, e.g., 2, 3, 5, 7, etc. Such numbers are calledprime numbers, and they play an important role, both in pure mathematics and its applications. The distribution of such prime numbers among all natural numbers does not follow any regular pattern, however the German mathematician G.F.B. Riemann (1826 - 1866) observed that the frequency of prime numbers is very closely related to the behavior of an elaborate function

$\zeta(s) = 1 + 1/2s + 1/3s + 1/4s + ...$

called the Riemann Zeta function. The Riemann hypothesis asserts that allinteresting solutions of the

equation

$\zeta(s) = 0$

lie on a certain vertical straight line. This has been checked for the first 1,500,000,000 solutions. A proof that

it is true for every interesting solution would shed light on many of the mysteries surrounding the distribution

of prime numbers. First of all in my opinion, if Riemann is true then the Zeta function must then be infinite because it is a continuing frequency of prime numbers. Simple graphs has identified this as true as seen here:

Proven infinite and true as having both obvious and non obvious values in the numerical function. Also known as below zero and above zero. All the shown graphs are plotted correctly, coordinates with zero, and are in support of the Riemann Hypothesis of a Zeta function.

A double graph is an example of the possibility of both obvious and non obvious numerical values. It also shows its in coordinace with a straight vertical line passing through zero.

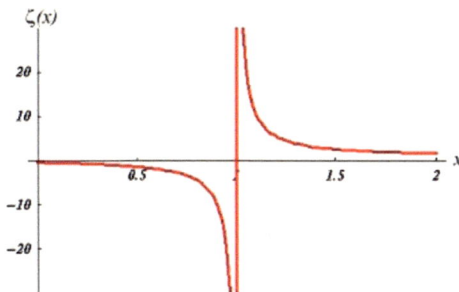

The Reinman Hypothesis states that the Zeta Function is vertical, this must be true logically, because the zeta function is infinite.

$\zeta(s) = 1 + 1\ 2^s + 1\ 3^s + 1\ 4^s + \ldots$

Means the the Zeta Function is infinite.

Zeta ends at 1+1 infinity's, which equals infinity o the ending functional value of Zeta & infinity Obvious zeros and Non-Obvious zeros are the two catagories for the b eginning and of the function. However, you can't estimate and end b ecause the Continuum of the functional equation. This means that my hypothesis of the Zeta function being infinite & correct, so Reinman's hypothesis is a meaningful representation of this data.

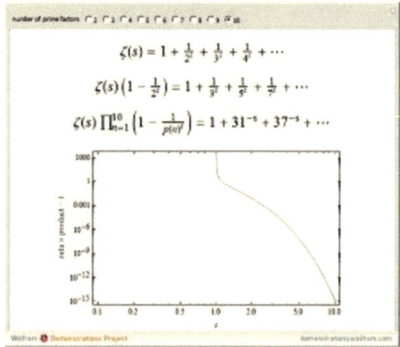

number of prime factors $\Gamma_2\,\Gamma_3\,\Gamma_4\,\Gamma_5\,\Gamma_6\,\Gamma_7\,\Gamma_8\,\Gamma_9\,\Gamma_{10}$

$$\zeta(s) = 1 + \tfrac{1}{2^s} + \tfrac{1}{3^s} + \tfrac{1}{4^s} + \cdots$$

$$\zeta(s)\left(1 - \tfrac{1}{2^s}\right) = 1 + \tfrac{1}{3^s} + \tfrac{1}{5^s} + \tfrac{1}{7^s} + \cdots$$

$$\zeta(s)\,\textstyle\prod_{n=1}^{10}\left(1 - \tfrac{1}{p(n)^s}\right) = 1 + 31^{-s} + 37^{-s} + \cdots$$

Look at the following above the prime factors keep continuing and will keep being in and out of the curve, meaning this is also another example of the idea that the Zeta Function has an infinite continuation, both datas, if correct as well as logically initiated, have proved my solution right.

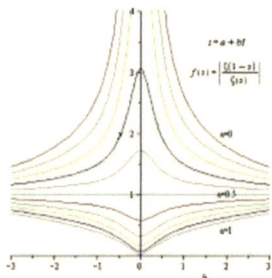

$z = a + bi$

$$f(z) = \left| \frac{\zeta(1-z)}{\zeta(z)} \right|$$

This equation will have the object representation increases in size meaning the graph will infinitely increase, this graph also proves my idea that the zeta function is infinite.

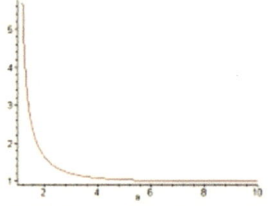

Same example of a graph continuing and same example that the Zeta is infinite, and proven that both the obvious and non obvious values will also keep going on.

If my idea that these graphs are in correspond with my idea then this would also correspond with the Im(s)=T argument where there exists an imaginary part between 0 and T.

$$\frac{1}{\pi}\text{Arg}(\Gamma(\tfrac{s}{2})\pi^{-s/2}s(s-1)/2) = \frac{T}{2\pi}\log\frac{T}{2\pi} - \frac{T}{2\pi} + 7/8 + O(1/T)$$

Sources:

- ^ a b R. E. Ladner "On the structure of polynomial time reducibility," Journal of the ACM, 22, pp. 151–171, 1975. Corollary 1.1. ACM site.

- ^ Cook, Stephen (1971). "The complexity of theorem proving procedures". *Proceedings of the Third Annual ACM Symposium on Theory of Computing*. pp. 151–158.

- ^ Lance Fortnow, *The status of the P versus NP problem*, Communications of the ACM 52 (2009), no. 9, pp. 78–86.doi:10.1145/1562164.1562186

- Artin, Emil (1924), "Quadratische Körper im Gebiete der höheren Kongruenzen. II. Analytischer Teil", *Mathematische Zeitschrift* **19** (1): 207–246, doi:10.1007/BF01181075

- Beurling, Arne (1955), "A closure problem related to the Riemann zeta-function", *Proceedings of the National Academy of Sciences of the United States of America* **41** (5): 312–314,doi:10.1073/pnas.41.5.312, MR 0070655

- Bohr, H.; Landau, E. (1914), "Ein Satz über Dirichletsche Reihen mit Anwendung auf die ζ-Funktion und die L-Funktionen", *Rendiconti del Circolo Matematico di Palermo* **37** (1): 269–272,doi:10.1007/BF03014823

- Bombieri, Enrico (2000), *The Riemann Hypothesis - official problem description* (PDF), Clay Mathematics Institute, retrieved 2008-10-25 Reprinted in (Borwein et al. 2008).